Interactions of Matter

by Christine Caputo

Table of Contents

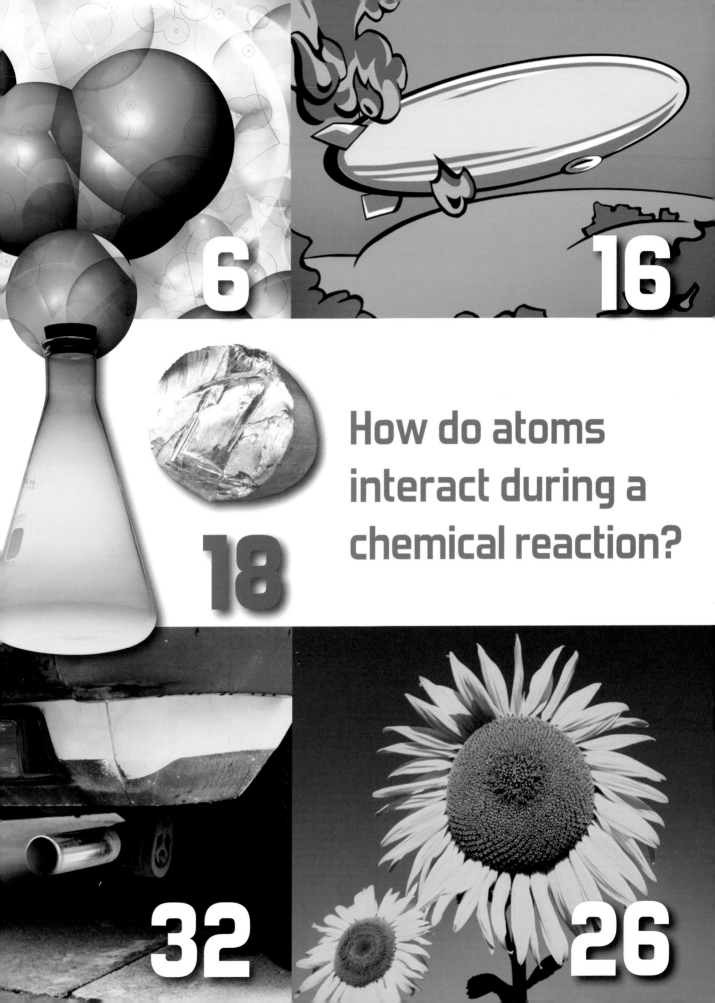

How do atoms interact during a chemical reaction?

The Mysterious Change in America's Symbol of Freedom

Standing proudly on Liberty Island in New York Harbor is the Statue of Liberty. France gave the statue to the United States as a gift to honor the 100-year anniversary of the Declaration of Independence. Today, the statue is a symbol of freedom to people around the world.

The French sculptor Frédéric Auguste Bartholdi (freh-duh-REEK au-GOOST bar-TAHL-dee) designed the Statue of Liberty. He developed the idea of a huge statue that would stand more than 46 meters (151 feet) tall. To build such an enormous structure, Bartholdi asked Alexandre Gustave Eiffel (ah-lek-ZAHN-der GOOS-tahv I-fel) for help. Eiffel had designed the now-famous Eiffel Tower in Paris. He developed a supporting frame for the statue made out of iron.

The iron frame was then covered in a layer of copper. Copper is a brownish metal. You see copper whenever you look at a penny. The copper layer on the Statue of Liberty is as thick as two pennies put together.

▲ This model shows what the face of the Statue of Liberty might have looked like when it was first made.

After it was built in France, the Statue of Liberty was separated into more than 300 pieces. The pieces were packed into crates and shipped to America. It took four months to put the statue back together. Then, on October 28, 1886, the Statue of Liberty was dedicated in front of thousands of people.

The people there that day in 1886 saw something different than you would see today. The giant statue was a shiny, reddish-brown color. Today, the Statue of Liberty is green! What caused the statue to turn from brown to green? The answer lies within the pages of this book!

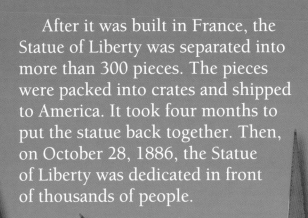

▲ Today, the Statue of Liberty is blue-green rather than brown.

The copper layer on the Statue of Liberty is as thick as two pennies put together.

Chemical Bonding

How and why do atoms combine?

Have you ever played a game in which you try to make as many different words as you can from the same letters? For example, what words can you make from the letters *a, m, o,* and *t*? Here are some possibilities.

You can put just four letters together in different ways to make many words. The words have different lengths, sounds, and meanings. Imagine how many words you can make from the entire alphabet of twenty-six letters!

Now think about all the different elements of matter in the world. Together the atoms of these elements combine to form all of the things that we can see, smell, taste, and touch. From the acorns on oak trees to the water in clouds, and from sand on a beach to glass in a window, matter is made up of atoms just as words are made up of letters. The types of atoms and the ways in which they are arranged determine the type of matter that results.

These particles of matter are held together by forces of attraction known as **chemical bonds**. Chemical bonding accounts for the many different types of substances that exist in the universe.

▲ If you look closely at this painting by the French artist Georges Seurat, you will discover that it is made up of many tiny dots of color. The dots come together to create a picture. In the universe, tiny particles held together by chemical bonds create the many types of matter.

Valence Electrons

Atoms are the building blocks of matter. Atoms are made up of protons, neutrons, and electrons. Protons and neutrons are locked in the nucleus, or center, of an atom. Electrons move at high speeds around the nucleus. Electrons may have different amounts of energy. A region in which electrons with the same energy are likely to be found is called an energy level. Electrons with the lowest amount of energy are closest to the nucleus. Electrons with more energy are farther from the nucleus.

The electrons in the highest energy level are known as **valence electrons**. These electrons are farthest from the nucleus and are held most loosely by the atom. The number of valence electrons in an atom determines if and how the atom will bond with other atoms. Chemical bonds form when valence electrons are either transferred or shared.

Different energy levels can hold only a certain number of electrons. If the outermost energy level of an atom does not have all of the electrons it can hold, it is not complete. An atom fills that level by bonding with another atom that is also incomplete. When atoms have filled their outermost energy level, they are stable. This means that they are less likely to change. For almost all atoms, a filled outermost energy level contains eight valence electrons. The exceptions are hydrogen and helium, whose atoms are stable with two valence electrons.

The electrons in the highest energy level are known as valence electrons.

▼ These models represent atoms. Each dot stands for one electron. The electrons in the outermost circle are valence electrons.

3 valence electrons

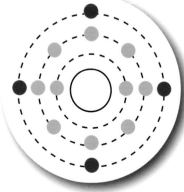

4 valence electrons

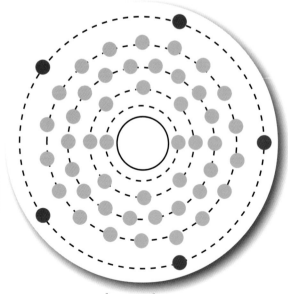

5 valence electrons

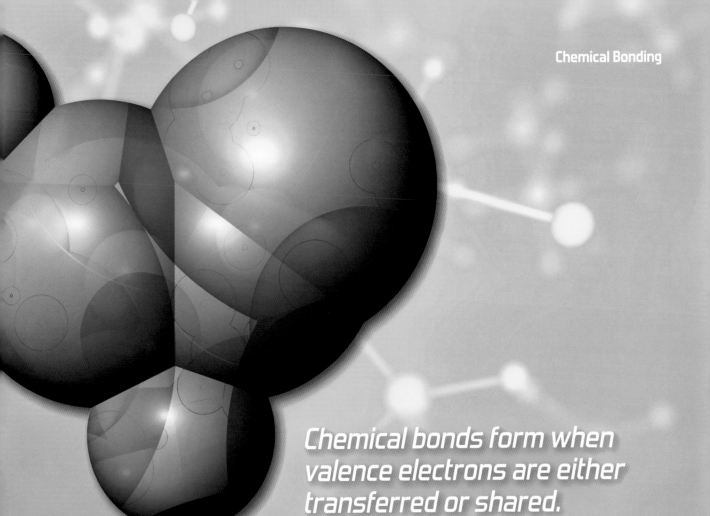

Chemical bonds form when valence electrons are either transferred or shared.

Science and Math

OCTET RULE

You can use mathematical prefixes to figure out the meanings of some unfamiliar words. Consider the word *octet*. The prefix is *oct-*, which means "eight." The octet rule states that most atoms are most stable when they have eight electrons in their outermost energy level.

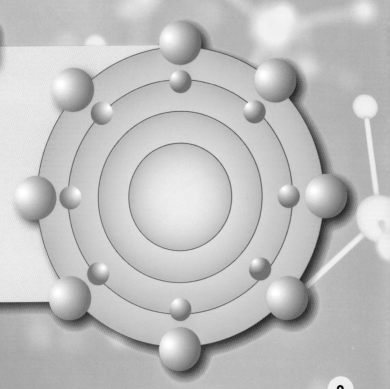

Ionic Bonds

Some atoms gain or lose electrons to become stable. This is how sodium chloride, which you know as table salt, is formed. A sodium atom has only one valence electron. Sodium atoms become stable by giving that electron away to a chlorine atom. A chlorine atom has seven valence electrons. It becomes stable by accepting one electron from a sodium atom.

Electrons have a negative electric charge. Protons have a positive electric charge. In an atom, the number of electrons equals the number of protons. The negative and positive charges are equal, which makes the atom electrically neutral. If an atom gains or loses electrons, a charged particle called an **ion** is produced. An atom becomes a positive ion if it loses electrons. It becomes a negative ion if it gains electrons. In sodium chloride, sodium is a positive ion because it lost its one valence electron. Chlorine is a negative ion because it gained one electron to have eight valence electrons.

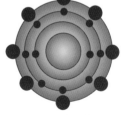

table salt

▼ The sodium atom gives one electron to the chlorine atom. Sodium becomes a positive ion and chlorine becomes a negative ion. An ionic bond joins the ions to form the compound sodium chloride, or table salt.

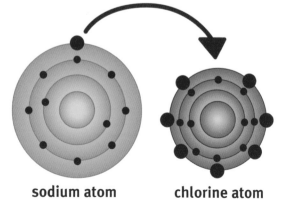
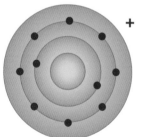

| sodium atom | chlorine atom | positive ion | negative ion |

A negative ion and a positive ion are attracted to each other by a force. This force is known as an **ionic bond**. The substance that forms from ionic bonding is called an ionic compound.

Ionic compounds exist in the form of crystals. Crystals have repeating arrangements of ions that give the crystals their particular shapes. The strong force of attraction between ions in a crystal gives ionic compounds certain properties, such as high melting points and boiling points. Ionic compounds also tend to be brittle solids at room temperature. They can break apart if they are hit with a hammer. When dissolved in water, ionic compounds can conduct electricity.

Science to Science

CHEMISTRY AND METEOROLOGY: SNOW

When water freezes into ice, it forms a crystal. An ice crystal or several crystals together make up a snowflake. A snowflake is not a frozen raindrop. When raindrops freeze, they form sleet. Snow crystals form when water vapor in the air condenses directly into ice. Unlike sleet, snow crystals have a regular, repeating pattern. As they grow, the patterns can become more complex.

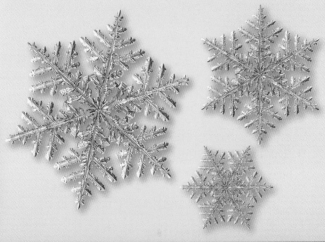

◀ The ions in sodium chloride (NaCl), or table salt, form a crystal that has a cubic shape.

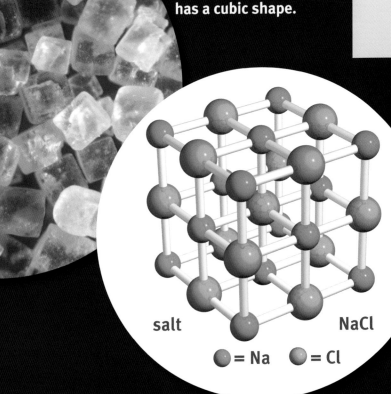

salt   NaCl

⬤ = Na   ⬤ = Cl

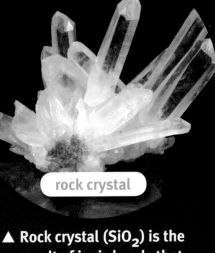

rock crystal

▲ Rock crystal (SiO_2) is the result of ionic bonds that form between silicon and oxygen molecules.

Covalent Bonds

Some atoms do not gain or lose electrons with ease. Instead, they form chemical bonds by sharing electrons. A **covalent bond** is the force of attraction between two atoms that share electrons. The diagram below shows how water forms: One oxygen atom shares electrons with two hydrogen atoms.

The combination of atoms formed by a covalent bond is called a **molecule**. A molecule is the smallest particle of a compound with covalent bonds that has all the properties of that compound. Some other molecular compounds that form covalent bonds are carbon monoxide, carbon dioxide, and ammonia. When compared with ionic compounds, molecular compounds generally have lower melting points and boiling points. Unlike ionic compounds, very few covalent compounds conduct electricity when dissolved in water.

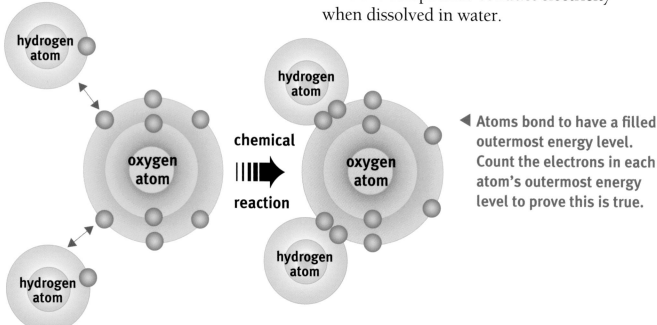

◀ Atoms bond to have a filled outermost energy level. Count the electrons in each atom's outermost energy level to prove this is true.

Science to Science

CHEMISTRY AND BIOLOGY: DNA

Many compounds essential to human survival are made up of molecules. The molecule that contains the genetic instructions for the development and functioning of almost all living things is known as DNA. DNA is often said to be "the blueprint of life." In all higher organisms, this molecule is made up of two strands that resemble the uprights of a ladder, but they are twisted in a spiral. The atoms in a DNA molecule (which include carbon and nitrogen) are held together by strong covalent bonds.

Metallic Bonds

Many elements are classified as metals. Jewelry is often made from the metals silver, gold, and platinum. Electrical wires are often made of the metal copper. The metal aluminum is used to make cans and foils. Metals are generally shiny solids at room temperature. They have high melting and boiling points. Metals can be drawn into wires or hammered into sheets. They are also good conductors of heat and electricity.

The atoms in a metal are tightly packed together in a specific arrangement. Most metals tend to lose their valence electrons to form positively charged ions. The valence electrons are free to drift among the positive ions. A **metallic bond** is the force of attraction between a positively charged metal ion and the electrons in a metal. The properties of metals result from the arrangement of positive ions and negative electrons. Because the electrons can move, the shape of the metal can be changed, and heat and electricity can flow through the metal.

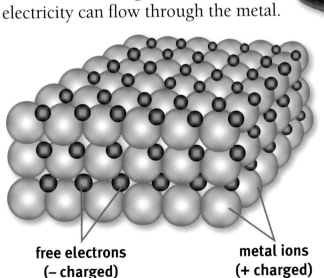

free electrons
(– charged)

metal ions
(+ charged)

▲ A metal is sometimes described as
ions floating in a sea of electrons.

Checkpoint
Talk It Over

Work in a small group to summarize the characteristics of the three types of chemical bonds in chart form. Make sure you highlight the differences among the types. Provide an example of each type in your chart. Present your chart to the class.

Metals have properties that make them useful in many ways.

▲ Unlike other common metals, mercury is a liquid at ordinary temperatures. Mercury expands when heated, which makes it perfect for use in thermometers.

▲ Like gold or silver, copper is highly malleable. That means when hot or cold, it can be bent and shaped without cracking. It can also be rolled into sheets as thin as 1/500 of an inch.

▲ Metals can be drawn into wires or hammered into sheets.

Summing Up

- Matter is made up of atoms that are held together by chemical bonds.

- Electrons in the outermost energy level of an atom are called valence electrons.

- Atoms are stable when their outermost energy levels are filled. For most atoms, this involves eight valence electrons.

- To become stable, some atoms gain or lose electrons through ionic bonding. An atom that gains or loses electrons forms an ion.

- A bond that allows atoms to share electrons is a covalent bond. A neutral group of covalently bonded atoms forms a molecule.

- Metallic bonding is the force of attraction between metal ions and electrons in a metal. The free movement of electrons gives metals their properties.

Putting It All Together

Choose from the research activities below. Work independently, in pairs, or in a small group. Share your responses with the class.

1 Work with a partner to create a hand-drawn or computerized display that explains the role of valence electrons in different types of chemical bonds. Be sure to show how and why atoms bond with each other.

2 Use items from around your home or classroom to make a model of a crystal. Present your model to the class and explain which type of bonding can result in this structure.

3 Choose a chemical compound that you enjoy every day. It could be something that you like to eat or drink. It could be something that you like to wear. Research the chemical formula for this compound and explain what types of bonds hold this compound together.

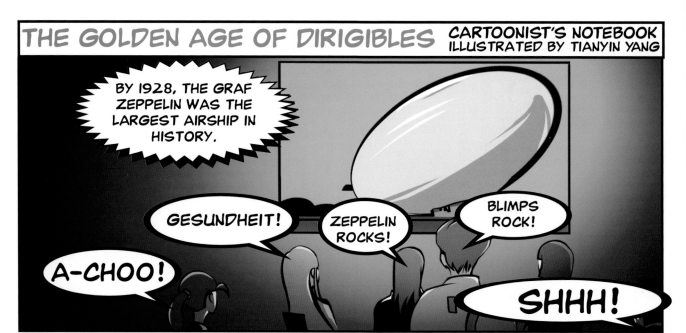

BY 1928, THE GRAF ZEPPELIN WAS THE LARGEST AIRSHIP IN HISTORY.

GESUNDHEIT!

ZEPPELIN ROCKS!

BLIMPS ROCK!

A-CHOO!

SHHH!

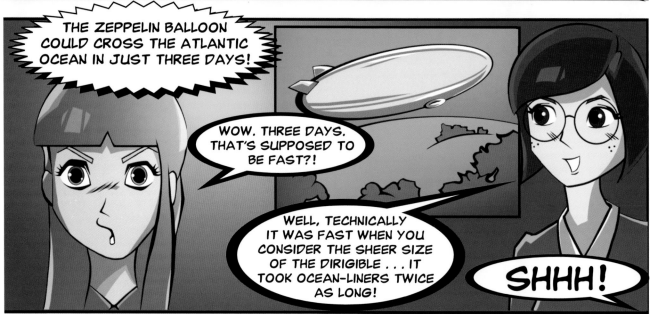

THE ZEPPELIN BALLOON COULD CROSS THE ATLANTIC OCEAN IN JUST THREE DAYS!

WOW. THREE DAYS. THAT'S SUPPOSED TO BE FAST?!

WELL, TECHNICALLY IT WAS FAST WHEN YOU CONSIDER THE SHEER SIZE OF THE DIRIGIBLE . . . IT TOOK OCEAN-LINERS TWICE AS LONG!

SHHH!

THE HINDENBURG WAS BUOYED WITH HYDROGEN, A HIGHLY FLAMMABLE GAS. . .

UH-OH. DO SEE WHERE THIS IS HEADED?

UH. WHERE?

ELECTRIC CURRENTS IN EARTH'S ATMOSPHERE IGNITED THE PAINT AND HYDROGEN . . .

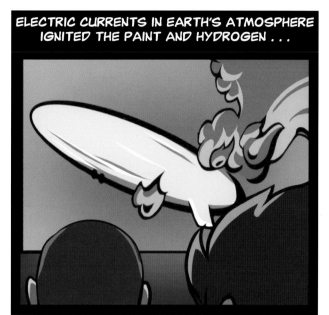

WITHIN SECONDS, THE HYDROGEN GAS EXPLODED AND THE GREAT *HINDENBURG* WAS GONE. . .

THEY PROBABLY SHOULD HAVE USED A NON-FLAMMABLE GAS LIKE HELIUM!

I CONCUR.

AND SO, ON MAY 6, 1937, THE GOLDEN AGE OF DIRIGIBLES CAME TO A TRAGIC END.

BUMMER!

I CONCUR.

SHHH!

WHEN THE HINDENBURG EXPLODED, A CHEMICAL REACTION TOOK PLACE.

HOW DO YOU KNOW?

WHAT ARE SOME OTHER TELL-TALE SIGNS THAT A CHEMICAL REACTION IS TAKING PLACE?

The Nature of Chemical Reactions

What happens during a chemical reaction and how can chemical reactions be described?

Suppose a chef needs to add a bit of table salt to a recipe. Would it make a difference if the chef grabbed brown sugar instead of salt? Of course it would, because the properties of brown sugar are different from those of table salt. A property is a characteristic.

Matter has two types of properties: physical properties and chemical properties. A physical property is a characteristic you can observe with your senses or determine through measurement. For example, you can see, feel, smell, and taste the difference between table salt and brown sugar. A chemical property is the ability of atoms to react and to change into a different substance. The ability to burn, rust, and tarnish are examples of chemical properties.

Physical Change

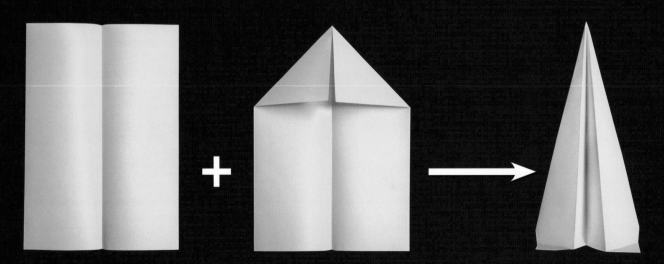

▲ Physical changes include cutting, bending, breaking, crushing, and reshaping matter. Although the matter may look different, it is still the same substance.

Changes in Matter

The properties of matter can change in different ways. When you break a sugar cube into pieces, you change the form of the sugar. You do not change the sugar into a different substance. Breaking the sugar cube apart is a physical change. A physical change is a change in the form or appearance of matter. Melting ice, tearing paper, sawing wood, and molding clay are examples of physical changes. In general, physical changes can be reversed.

Unlike a physical change, a chemical change produces one or more new substances. A chemical change is the result of a chemical reaction. During a **chemical reaction**, atoms of one substance rearrange to form a new substance. The new substance has different physical and chemical properties. The substances that enter into a chemical reaction are known as **reactants**. The substances that result from a chemical reaction are known as **products**. A chemical reaction cannot be reversed by physical processes. The products cannot be changed back into the reactants by physical processes.

Chemical Change

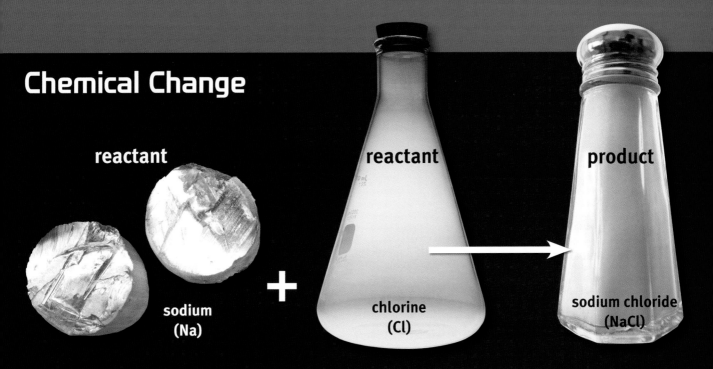

reactant

sodium
(Na)

+

reactant

chlorine
(Cl)

product

sodium chloride
(NaCl)

▲ Sodium is a soft, silver metal that reacts violently with water. Chlorine is a greenish-yellow gas that is dangerous to breathe. They combine to form table salt, which is made up of white crystals that are safe to eat.

Evidence of Chemical Reactions

With an almost deafening noise and bright flames, dynamite explodes in an empty building. The walls of the building collapse and the building is demolished. The explosion of dynamite is a dramatic example of a chemical reaction. Not all chemical reactions are quite as obvious, however. Sometimes, you need to look for clues to figure out if a chemical reaction has occurred. Although you cannot reach a conclusion from a single clue, several clues together can help you decide if a chemical reaction has occurred.

PRECIPITATE ▶

Some chemical reactions between liquids result in the production of a solid that usually sinks to the bottom. The solid is called a precipitate.

◀ GAS CHANGE

In some chemical reactions, a gas is formed. You can see the gas as bubbles that rise up through a liquid. Note: Not all bubbles mean a chemical reaction has happened.

◀ ENERGY CHANGE

Some chemical reactions occur when the reactants absorb energy. Others involve a release of energy. Therefore, an increase or decrease in energy suggests that a chemical reaction has occurred.

COLOR CHANGE ▶

A change in color occurs in many types of chemical reactions. Changes in smell and taste are often produced as well, but you should never taste a specimen.

Everyday Science

RIPENING

As fruits ripen, they become sweeter, softer, and less green. Ripening involves a chemical reaction during which starch in the fruit is changed into sugar. When a fruit begins to ripen, or is damaged, it releases a gas called ethylene. This gas speeds up the ripening process in other fruits. If you want to make fruits ripen faster, you can store them with fruits that are already ripe. You can also seal them in a bag to trap the ethylene gas.

Hands-On Science

CHANGING YOUR LOOSE CHANGE

A chemical change in matter produces one or more new substances. These substances have physical and chemical properties different from the original substance. In this experiment, you will investigate a chemical change. What evidence indicates a change has occurred? How can you change the rate at which a chemical change takes place?

Time: several days

Materials:
- 4 copper pennies (before 1981 if possible)
- 4 clear plastic cups
- marker
- salt
- vinegar
- water

STEP 1: Place one penny in each cup.

STEP 2: Use the marker to label each of the four cups as one of the following:

- vinegar
- vinegar and salt
- water
- water and salt

STEP 3: Sprinkle a teaspoon of salt over the pennies in the cups with "Salt" in their labels.

STEP 4: Pour enough vinegar to cover the pennies in the cups with "Vinegar" in their labels.

STEP 5: Pour enough water to cover the pennies in the cups with "Water" in their labels.

STEP 6: Observe how the pennies look each day for one week. Record your observations in a data table similar to the one here.

STEP 7: Compare your observations for each penny. Which samples yielded a chemical change? What evidence is there that a chemical change has occurred? How do vinegar and salt affect the chemical change you observed?

Day	1	2	3	4	5	6	7
Vinegar only							
Vinegar and salt							
Water only							
Water and salt							

Chemical Equations

If you were to describe in words everything that occurs during a chemical reaction, you would need lots of time and lots of paper. Other people might not understand your description. This is because they might use different words to describe the same event. Scientists have developed a shorthand for describing chemical reactions. This quick method of description is understood by all scientists. A **chemical equation** is a description of a chemical reaction using chemical formulas and symbols.

A **chemical formula** shows the ratio of atoms in a compound. When you write a chemical formula, you use the chemical symbols for the elements in the compound. The chemical formula for carbon dioxide is CO_2. The C represents a carbon atom and the O represents an oxygen atom. The 2 is a subscript. A subscript is a small number written to the bottom right of a chemical symbol. It indicates the number of atoms of that element in each molecule of the compound. In CO_2, the 2 tells you that there are 2 atoms of oxygen in each molecule of carbon dioxide. If there is no subscript written, the number of atoms is 1.

The chemical equation below describes the formation of carbon dioxide from carbon and oxygen.

$$C + O_2 \rightarrow CO_2$$

▲ This chemical equation shows that carbon combines with oxygen to produce carbon dioxide.

Balancing Chemical Equations

The **law of conservation of mass** (also known as the law of conservation of matter) states that atoms are neither created nor destroyed during a chemical reaction. This means that all of the atoms of the reactants are present in the products. The number of atoms of each element on the left side of the equation is equal to that of the right side of the equation. To show this, an equation must be balanced.

Balancing an equation is similar to solving a puzzle or a math problem. Consider the reaction in which nitrogen gas (N_2) and hydrogen gas (H_2) form ammonia (NH_3). This is the unbalanced equation:

$$N_2 + H_2 \rightarrow NH_3$$

Counting atoms shows:

$$N_2 + H_2 \rightarrow NH_3$$
$$N = 2 \qquad N = 1$$
$$H = 2 \qquad H = 3$$

The number of atoms of each element is not equal on both sides of the equation. To make them equal, **coefficients** (koh-ee-FIH-shunts) must be used. A coefficient is a number written in front of a chemical symbol or formula. The coefficient is multiplied by the subscript of each element in the formula to determine the number of atoms.

Starting with the nitrogen atoms, a coefficient of 2 is written in front of th product and the atoms are again count

$$N_2 + H_2 \rightarrow 2NH_3$$
$$N = 2 \qquad N = 2$$
$$H = 2 \qquad H = 6$$

Nitrogen is balanced—2 atoms on each side of the arrow—but hydrogen not. There are 2 atoms in the reactants and 6 atoms in the products. Since 6 is 2 x 3, a coefficient of 3 is written in fro of H_2. The chemical equation is now correctly balanced.

$$N_2 + 3H_2 \rightarrow 2NH_3$$
$$N = 2 \qquad N = 2$$
$$H = 6 \qquad H = 6$$

Checkpoint ✔

Reread

Reread the steps in balancing a chemical equation. Then test your skill by balancing the following equation:

$$Fe + Cl_2 \rightarrow FeCl_3$$

Answer: $2Fe + 3Cl_2 \rightarrow 2FeCl_3$

Hands-On Science: CONSERVING MATTER

During a chemical reaction, new substances are formed. These products have physical and chemical properties different from those of the reactants. Although new substances are produced, matter is neither created nor destroyed. It is conserved. How can you demonstrate the law of conservation of matter?

TIME: 45 minutes (plus 5–10 minutes each week for 2 weeks)

MATERIALS: a piece of steel wool about the size of an egg

250-mL beaker   white vinegar

250-mL flask   balloon

balance

STEP 1:   Place the steel wool in the beaker and add white vinegar until the steel wool is covered. Soak for about five minutes.

STEP 2:   Remove the steel wool and wring out any excess vinegar.

STEP 3:   Place the steel wool in the flask. Stretch the balloon over the opening.

STEP 4:   Use the balance to find the mass of the flask with the steel wool and the balloon. Record your measurement.

STEP 5:   Set the flask aside for one week.

STEP 6:   Use the remaining lab period to answer the following questions:
- What is steel wool composed of?
- What is present in the flask besides iron?
- What are the major gases found in air?
- What observations might lead you to believe that the iron was chemically reacting with O_2?
- Write a word equation for the reaction between iron and oxygen.

STEP 7:   At the end of 1 week, observe the steel. Find the mass of the flask with the steel wool and the balloon again. What did you observe about the balloon this time? Why do you think this happened? Record your measurements and observations in a table or chart.

STEP 8:   At the end of 2 weeks, repeat step 7. Then compare the masses you found. Draw a conclusion about what this tells us about chemical reactions.

Science to Science

CHEMISTRY AND GEOLOGY: LIMESTONE

Huge underground caverns that form in limestone rock are the results of chemical reactions. Water that seeps into the ground combines chemically with carbon dioxide to form a substance called carbonic acid. The carbonic acid dissolves calcite, which is the main mineral in limestone. Over many years, these reactions can carve out large spaces in the underground rock.

Summing Up

- A physical property of matter is a characteristic that can be observed with the senses. A physical change affects only the form or appearance of matter.

- A chemical property of matter is an ability to undergo change. During a chemical change, or chemical reaction, reactants change into products.

- According to the law of conservation of matter, no atoms are created or destroyed during this process.

- Evidence that suggests that a chemical reaction has occurred includes formation of a gas, formation of a precipitate, change in color, and change in energy.

- A chemical reaction can be described by a chemical equation.

- A balanced chemical equation uses symbols and chemical formulas to show that matter is conserved.

Putting It All Together

Choose from the research activities below. Work independently, in pairs, or in a small group. Share your responses with the class.

1 Research a common chemical change. Then, on a sheet of poster board, show the chemical reaction that causes this change in words, symbols, and pictures. (Be sure to label the reactants and products.)

2 Use pencil erasers and toothpicks to model the unbalanced equation $H_2 + O_2 \rightarrow H_2O$. Then use more erasers to balance the equation.

3 Create a play, movie, or animated cartoon showing the difference between physical and chemical changes. You can focus your presentation on one particular type of matter and the different physical and chemical changes this type of matter might experience, or you can use a variety of examples.

Types of Chemical Reactions

What are the different types of chemical reactions?

There are millions of different compounds. That means that there are millions of different chemical reactions that form them. Some chemical reactions are loud and explosive. Others are so quiet that you hardly notice them. Some reactions occur instantly, whereas others occur slowly over time. To bring order to the many reactions, scientists classify them into major groups:

- synthesis reactions
- decomposition reactions
- single-replacement reactions
- double-replacement reactions
- combustion reactions

The Root of the Meaning: The word **SYNTHESIZE** comes from the Greek word *syntithenai*, **which means "to put together."**

Sodium combines with chlorine in the synthesis reaction that forms sodium chloride, or table salt. ▶

Rusting is a common synthesis reaction in which iron combines with oxygen. ▶

Synthesis Reactions

During a **synthesis reaction**, two or more substances combine to form a single compound. The word *synthesize* means "to put together." The substances put together can be elements, compounds, or both.

One of the most important chemical changes on Earth is the synthesis reaction that takes place in green plants. In this reaction, two substances combine to form two new substances. During photosynthesis, green plants combine water and carbon dioxide in the presence of sunlight to produce glucose (a type of sugar) and oxygen.

$$6H_2O + 6CO_2 \rightarrow C_6H_{12}O_6 + 6O_2$$

◄ Photosynthesis is a synthesis reaction.

The formation of water is a synthesis reaction in which hydrogen combines with oxygen. ▶

Decomposition Reactions

In a **decomposition reaction**, a single compound breaks down to form two or more simpler substances. Many decomposition reactions occur when electricity or heat is applied to a substance.

$$2H_2O \rightarrow 2H_2 + O_2$$

▲ The decomposition of water occurs when an electric current is passed through the water. The products are hydrogen and oxygen. This decomposition reaction is the opposite of the synthesis reaction in which water is formed.

$$2HgO \rightarrow 2Hg + O_2$$

▲ Oxygen was first discovered in 1774 by Joseph Priestly as a product of a decomposition reaction.

▲ Through a complex series of decomposition reactions, mushrooms and other decomposers break down the bodies of dead and decaying organisms. Through decomposition reactions, the decomposers obtain food and return nutrients to the soil.

▲ A decomposition reaction causes air bags to inflate in a car that has been involved in an accident.

Single-Replacement Reactions

In a **single-replacement reaction**, one element takes the place of another element in a compound. This type of reaction can be used to separate one substance from another, such as a metal from a compound. As a result, these reactions have many important uses in industry.

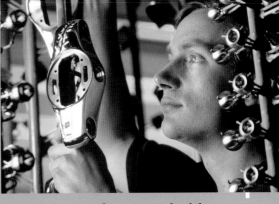

$$Cu + 2AgNO_3 \rightarrow Cu(NO_3)_2 + 2Ag$$

▲ Metals are often coated with a thin layer of another metal in a process known as electroplating. This type of single-replacement reaction can protect objects from corrosion or give a gold coating to less expensive metals.

▲ In this single-replacement reaction, copper replaces silver. The copper seems to disappear and a silvery-white metal appears in a blue solution.

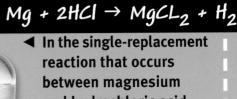

$$Mg + 2HCl \rightarrow MgCL_2 + H_2$$

◄ In the single-replacement reaction that occurs between magnesium and hydrochloric acid, magnesium takes the place of hydrogen.

HYDROCHLORIC ACID HCl

Double-Replacement Reactions

In a **double-replacement reaction**, parts of two compounds switch places to form new substances. The compounds are usually ionic compounds. It is the ions that switch places during the reaction.

Antacids take part in a double-replacement reaction with acid in the stomach. Once the acid is neutralized, stomach discomfort usually disappears. ▶

◄ A type of double-replacement reaction has been used since ancient times to make soap.

Combustion Reactions

In a **combustion reaction**, a substance combines with oxygen. Substances often burn during combustion reactions, producing heat and light. Fuels such as wood, gasoline, propane, and natural gas burn during combustion reactions. Carbon dioxide and water are the products of most combustion reactions.

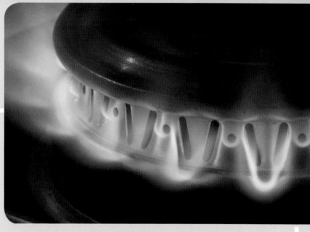

▲ Millions of people burn propane in furnaces, water heaters, outdoor grills, fireplaces, clothes dryers, and stoves. Propane burns in a combustion reaction.

Cellular respiration is a combustion reaction in which the cells of living things use oxygen to release the energy stored in food. This type of reaction is considered an example of slow combustion because it occurs at low temperatures. ▶

Science to Science

EARTH SCIENCE: THE CARBON CYCLE

One of the most abundant elements in living things is carbon. Carbon moves between the living and nonliving environment through the carbon cycle. One way that carbon is released into the environment is through the combustion, or burning, of fossil fuels like coal, oil, and gas. Fossil fuels formed from the remains of living things over millions of years. The carbon stored in those living things is now stored in the fossil fuels. When fossil fuels are burned, the carbon is released in the form of carbon dioxide. If too much carbon dioxide is released, it can upset the natural balance of the carbon cycle and alter Earth's atmosphere.

The Root of the Meaning:

The word

COMBUSTION

comes from the Latin *combustus*, which means "to burn up."

▲ Pockets of natural gas are found around sources of oil in the ground. The flames above an oil rig result from the combustion of natural gas that cannot be contained.

Summing Up

- Two or more reactants combine to form a single product in a synthesis reaction.

- During a decomposition reaction, a single reactant breaks down to form two or more products.

- An element serving as a reactant takes the place of another element in a compound during a single-replacement reaction.

- Parts of two compounds switch places to form new compounds in a double-replacement reaction.

- During a combustion reaction, a substance combines with oxygen, or burns.

Putting It All Together

Choose from the research activities below. Work independently, in pairs, or in a small group. Share your responses with the class.

1 Imagine a dance floor with lots of people dancing. Use this situation to describe single-replacement and double-replacement reactions in terms of the dancers. Illustrate or act out your description for your classmates.

2 Write each of the following chemical equations on a sheet of paper. First balance the equations. Classify each reaction. Then use colored markers to show how atoms or ions separated, combined, or switched places during each reaction.

$Cu + AgNO_3 \rightarrow Ag + Cu(NO_3)_2$ $K + Cl_2 \rightarrow KCl$

$C_2H_6 + O_2 \rightarrow CO_2 + H_2O$ $MgCl_2 \rightarrow Mg + Cl_2$

$FeS + HCl \rightarrow FeCl_2 + H_2S$

3 Create a poster, collage, or chart comparing and contrasting each of the reactions described in this chapter. Be sure to include examples of each type of reaction.

The Energy of Chemical Reactions

How is energy related to chemical reactions?

An important component of any chemical reaction is energy. Energy is stored in the bonds that hold atoms together. For example, there is energy stored between the oxygen atom and the hydrogen atoms in a water molecule. During a chemical reaction, chemical bonds are broken and new ones are formed as atoms are rearranged. It takes energy to break and form chemical bonds. In science, energy is the ability to do work or cause change.

The explosion of fireworks is an exothermic reaction. Another exothermic reaction is cellular respiration, the process by which this skateboarder releases energy stored in the foods he eats. ▶

The Root of the Meaning:
The word

EXOTHERMIC

comes from the Greek prefix **exo-**, meaning "**outside**," and the Greek word **thermein**, meaning "**heat**." An exothermic reaction releases heat to the outside. The word **endothermic** comes from the Greek prefix **endo-**, meaning "**inside**," and the Greek word **thermein**, meaning "**heat**." An endothermic reaction takes heat inside.

Exothermic Reactions

A chemical reaction that releases energy is called an **exothermic reaction**. All combustion reactions are exothermic. You can see the light and feel the warmth of a fire, for example. Light and heat are forms of energy. The process by which simple food nutrients are combined with oxygen to release energy is an exothermic reaction. During digestion, food is broken down into simpler substances and carried to all the cells of the body. Energy stored in food is released for the body to use during cellular respiration.

Endothermic Reactions

A chemical reaction that absorbs energy is called an **endothermic reaction**. Decomposition reactions that need either heat or electricity are endothermic because the reactant requires energy to change into the products.

Photosynthesis is an example of an endothermic reaction. Photosynthesis is the process by which plants make food. During photosynthesis, plants use the energy of sunlight to change carbon dioxide and water into sugar and oxygen.

Have you ever used a cold pack to prevent swelling at the site of an injury? A chemical cold pack works on the basis of an endothermic reaction. When the cold pack is bent or shaken, the reactants—usually water and ammonium nitrate—combine. As they react, they absorb heat from their surroundings. If you put the ice pack on your knee, your knee will feel cold because heat flows from your body into the cold pack.

▼ The chemical reactions that occur when foods cook or bake are endothermic reactions. The reaction that makes a cold pack work is also endothermic.

Energy Diagrams

Changes in energy always accompany chemical reactions. To get a "snapshot" of those changes, scientists use energy diagrams. Look at the two energy diagrams. For an exothermic reaction, you can see that the energy of the products is lower than the energy of the reactants. That is because energy is released as heat. For an endothermic reaction, the energy of the products is greater than the energy of the reactants. That is because energy is absorbed.

All chemical reactions—whether they are exothermic or endothermic—need a certain amount of energy to get started. Look at the energy diagrams again. Do you see that both diagrams have a peak to which the reactants must "climb" before they can form products? This peak represents the minimum amount of energy needed to start a chemical reaction.

Conservation of Energy

Although energy changes during a chemical reaction, it is neither created nor destroyed. Energy lost by the substances in a chemical reaction is released into the environment. Energy taken in by the substances in a chemical reaction is taken out of the environment. The overall energy of the substances in the chemical reaction and the environment stays the same.

This is known as the **law of conservation of energy**. This law states that energy is neither created nor destroyed during chemical reactions.

Checkpoint
Read More About It

The energy stored in the bonds that hold matter together is a type of potential energy. Find out about potential energy and kinetic energy. Find out what they have in common and how they are different.

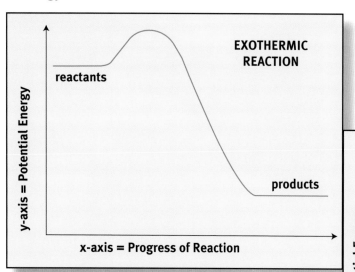

▲ Energy diagrams can be used to compare the energy of the reactants and products in chemical reactions.

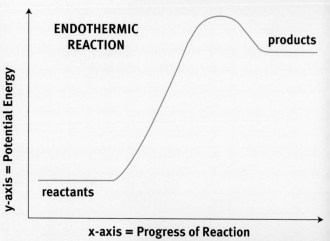

Everyday Science

RISING DOUGH

If you have ever helped someone make bread, you know that the dough needs to be warm in order to rise. Bread dough contains living things called yeasts. When they are warm enough, yeasts digest sugars in the dough. During digestion, the yeasts release carbon dioxide gas. Bubbles of carbon dioxide cause the dough to rise.

Rates of Reaction

Some chemical reactions occur quickly and others occur more slowly. Put a match to a pile of sawdust, and burning takes place almost instantly. Put the match to a thick log, and burning might take hours. The speed at which a chemical reaction takes place, or how quickly reactants turn into products, is known as the rate of reaction.

One factor that affects the rate of a reaction is the movement of the particles that make up matter. Particles of matter are in constant motion. As they move, they collide with each other. These collisions enable the particles to interact: old bonds may be broken and new bonds may be formed.

In order for a chemical reaction to occur, the particles must collide with the right amount of energy. Remember that particles need to have a minimum amount of energy to react—the peak in the energy diagrams. The more energy the particles have, the faster they will move and the more frequently they will collide. Increasing the number and energy of collisions between particles increases the rate of a reaction. Several factors affect the collisions between particles.

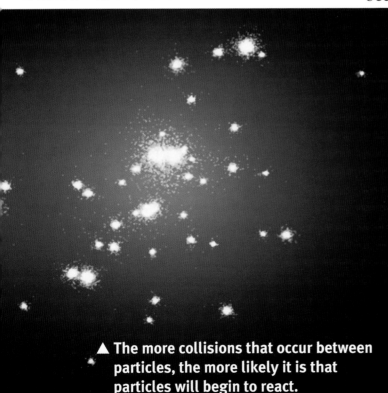

▲ The more collisions that occur between particles, the more likely it is that particles will begin to react.

TEMPERATURE

The temperature of a substance is a measure of the average motion of its particles. Raising the temperature of a substance causes the particles to move faster. When particles move faster, they collide more often with more energy. So increasing the temperature increases the rate of reaction. The opposite is also true. Lowering the temperature decreases the rate of reaction. Putting food in a refrigerator slows the rate of the reaction in which food spoils.

▲ You can see the difference in the amount of light, which indicates how fast the reaction is occurring.

CONCENTRATION

The concentration of a substance describes the number of particles in a given volume. Increasing the concentration of a substance means there are more particles to collide with each other. So increasing the concentration increases the reaction rate. Have you ever seen someone fan glowing coals to get a barbecue burning faster? Blowing air onto a small fire increases the concentration of air, and therefore oxygen, needed for the combustion reaction to occur.

Increasing the temperature increases the rate of reaction. Lowering the temperature decreases the rate of reaction.

SURFACE AREA

The surface area of a substance is how much of the substance is exposed. An increase in surface area increases the collisions between particles, and thus the rate of reaction. Suppose a solid is placed in a liquid. The particles of the liquid can collide only with the surface of the solid. If the solid is broken into pieces, the particles of the liquid can collide with more of the surface.

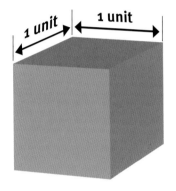

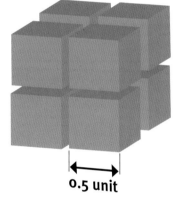

◄ The diagram shows that breaking the cube into pieces increases the surface area. You can also see that increasing the surface area of the steel wool increases the reaction rate.

Total surface area of cube:
1 unit x 1 unit x 6 faces =
6 square units

Total surface area of all cubes:
0.5 unit x 0.5 unit x 48 faces =
12 square units

CAREER

MEDICAL LABORATORY TECHNICIAN

A medical laboratory technician plays an important role in detecting and helping diagnose disease. Among other things, people in this field study chemical reactions in the human body. They investigate rates of reaction and factors affecting the rates.

Medical laboratory technicians usually have a bachelor's degree with a major in medical technology or in one of the life sciences. They also need on-the-job training and additional training related to their specific tasks. Medical laboratory technicians usually work in hospitals, but they can also be employed in research laboratories and government agencies.

Catalysts

Did you know that if it were not for certain substances in your saliva, it could take weeks for you to digest a simple cracker? Many chemical reactions that occur slowly cannot be sped up enough by increasing temperature, surface area, or concentration. However, they can be made to go faster by the addition of a catalyst. A **catalyst** is a substance that speeds up the rate of a chemical reaction but is not itself changed by the reaction. The catalyst is the same at the end of the reaction as it was in the beginning. A catalyst affects a chemical reaction by lowering the minimum amount of energy needed to start the reaction.

Catalysts are often used in industrial processes, such as refining petroleum and making plastic, vinyl, ammonia, and margarine. Because a small amount of catalyst can be used over and over again, catalysts save energy and reduce costs.

An **inhibitor** is a substance that works in the opposite way compared to a catalyst. An inhibitor slows the rate of reaction. Inhibitors are useful because some chemical reactions can occur too quickly or can keep going without stopping. Inhibitors make it possible to control or stop such reactions. Inhibitors are used in industrial processes, medicinal products, and food products. For example, many food preservatives inhibit, or slow, the natural decomposition reactions in food to preserve them longer.

▲ A catalyst lowers the minimum amount of energy needed for a chemical reaction to start. Catalysts in a car's catalytic converter affect the chemical reactions that produce exhaust. They reduce the amount of harmful gases released into the atmosphere by helping convert these pollutants into less harmful emissions.

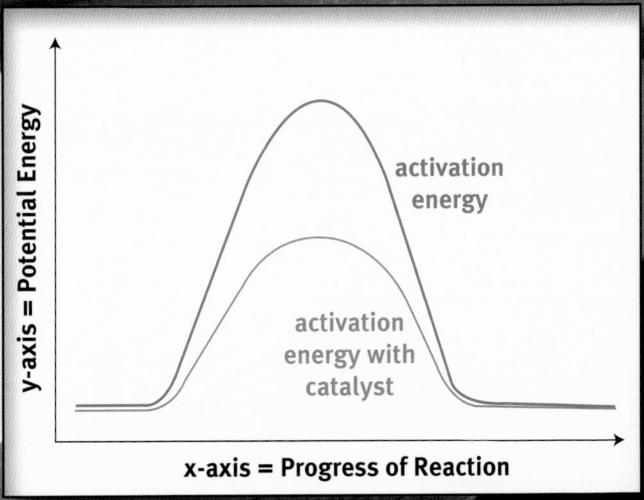

y-axis = Potential Energy

activation energy

activation energy with catalyst

x-axis = Progress of Reaction

Science to Science

CHEMISTRY AND BIOLOGY: ENZYMES

Catalysts are essential in the human body. Without them, many of the chemical reactions needed for life would proceed much too slowly. Natural catalysts called enzymes speed up some reaction rates by as much as one billion times. Enzymes are usually specific to one type of chemical reaction, so the human body contains many different types of enzymes for the various chemical reactions that occur within the body.

Hands-On Science
REACTION RATE

Reaction rate, or the speed at which a chemical reaction takes place, can be influenced by several factors, two of which are temperature and surface area. How do changes in these two factors affect the rate of reaction?

Time: 40 minutes

Materials: 500-mL beaker
3 effervescent tablets (such as antacid tablets)
200 mL room-temperature water
200 mL ice water
200 mL warm water
stopwatch
paper towels

STEP 1: Break each tablet into four equal pieces.

STEP 2: On a sheet of paper, create a data table similar to the one below.

STEP 3: Add 200 mL of room-temperature water to the beaker.

STEP 4: Drop a one-quarter tablet piece into the beaker. Start your stopwatch. Stop the stopwatch when the tablet is completely dissolved. Record the time in the data table.

STEP 5: Empty the contents of the beaker into a sink. Clean out the beaker with a paper towel.

STEP 6: Break a one-quarter tablet piece into smaller pieces. Then repeat Steps 3–5.

STEP 7: Grind one of the small pieces into a powder. Then repeat Steps 3–5.

STEP 8: Repeat Steps 3–7, but this time use ice water. Be sure to record your data.

STEP 9: Repeat Steps 3–7, but this time use warm water. Be sure to record your data.

STEP 10: Analyze your data. Draw a conclusion about how temperature and surface area affect the rate of reaction.

Room Temperature		Ice Water		Warm Water	
Tablet Size	Dissolving Times	Tablet Size	Dissolving Times	Tablet Size	Dissolving Times
Quarter		Quarter		Quarter	
Pieces		Pieces		Pieces	
Powder		Powder		Powder	

Summing Up

- Energy is released during exothermic reactions and absorbed during endothermic reactions.

- All chemical reactions require a minimum amount of energy in order to start. The law of conservation of energy states that energy is neither created nor destroyed during chemical reactions.

- The energy changes that occur during a chemical reaction can be summarized in an energy diagram.

- In order for a chemical reaction to occur, particles of reactants must collide with enough energy and frequency to interact.

- The number of collisions between particles and the amount of energy they have affect the rate of reaction, or the speed with which reactants become products.

- Four factors affect the rate of reaction: temperature, concentration, surface area, and the use of catalysts and inhibitors.

Putting It All Together

Choose one of the research activities below. Work independently, in pairs, or in a small group. Share your responses with the class.

1 Compare endothermic and exothermic reactions. Find or draw a picture representing an example of each type of reaction.

2 Create a diagram illustrating how temperature, concentration, and surface area affect the rate of reaction. Be creative in representing the particles of matter involved in the reaction.

3 Choose one of the factors affecting the rate of chemical reactions such as temperature or surface area and create an experiment that tests the accepted theory or idea about this factor.

Looking a Little Green

Chemical reactions are occurring all around you. In different ways, reactants change into products. You can often see evidence of a chemical reaction having occurred. Sometimes it's in the form of a change in color. Does that help you figure out what happened to the Statue of Liberty?

The change in color from brown to green suggests that the Statue of Liberty underwent a chemical reaction. The copper in the statue reacted with oxygen in the air to produce copper oxide. Copper and oxygen are the reactants and copper oxide is the product. The green coating that forms in this way is known as a patina. You formed a patina on pennies when you completed the experiment in Chapter 2.

In many places on the Statue of Liberty, the green patina is as thick as the copper that remains beneath it. The properties of the patina are different from the properties of the copper. In fact, the patina protects the remaining copper from being exposed to air. This prevents the copper from taking part in additional chemical reactions. In this way, the product of the chemical reaction that changed the statue's color actually protects the statue.

How to Write a Lab Report

Communicating the results of an experiment is an important part of the scientific method, and writing a lab report is one way to do so. A lab report must allow others to repeat your tests, review your conclusions, and build on your results, so your lab report must be as complete and accurate as possible. You will need to supply your preliminary observations, the information about the experiment you conducted, and the data you collected. Here are the steps to follow when writing a lab report.

Introduce your question or problem.

Describe the question or problem that your experiment will answer. Be sure to state it as a question. Explain why this question is important. What observations got you interested in the question? Mention any research that you have done on the topic. To write the report in the third person, avoid using words like "I" and "we."

State your hypothesis.

A hypothesis is a prediction that should be stated as an answer to the question. Based on all that you know about your topic, state what you expect to find.

List the materials you use.

Materials also include equipment. Be specific about amounts whenever such information is important.

Indicate safety precautions.

Doing experiments is often exciting. Doing them should always be safe. Be aware of any safety precautions you observe and indicate them.

Describe the procedure.

List all of the steps that you follow in the correct order.

Present your results.

You will want to present all of the data that you collect. Your data can be both quantitative and qualitative. Use tables, charts, and graphs to present the quantitative data. Describe qualitative observations.

Write your analysis and conclusion.

Use the data you collect to accept or reject your hypothesis. Explain the conclusion that you reach after completing the experiment. Restate your hypothesis and indicate if your experimental results support or fail to support your hypothesis. Think about the results of your experiment. What could you do differently next time? Is there a way to improve your experiment? Did your experiment generate any new questions?

Cite any references used.

You may have used books, magazines, newspapers, interviews, or the Internet to research your topic. Be sure to list the author, title, date, and place of publication or Web address for each source.

SURFACE AREA AND REACTION RATE

Question or Problem

How does surface area affect the rate of reaction?

Hypothesis

If surface area increases the rate of a reaction, a finely ground powder will have a faster reaction rate than larger pieces of antacid.

Materials

- 500-mL beaker
- stopwatch
- 9 effervescent tablets (such as antacid tablets)
- 2 liters room-temperature water
- paper towels

Safety Precautions

Be sure to wear safety goggles to protect your eyes during the experiment.

Procedure

STEP 1: Break each tablet into four equal pieces.

STEP 2: On a sheet of paper, create a data table similar to the one below.

STEP 3: Add 200 mL of room-temperature water to the beaker.

STEP 4: Drop a quarter-size tablet piece into the beaker. Start your stopwatch. Stop the stopwatch when the tablet is completely dissolved. Record the time in the data table.

STEP 5: Empty the contents of the beaker into a sink. Clean out the beaker and dry it with a paper towel.

STEP 6: Break a quarter-size tablet piece into smaller pieces. Then repeat Steps 3-5.

STEP 7: Grind one of the small pieces into a powder. Then repeat Steps 3-5.

STEP 8: Repeat the experiment two more times and average your results.

Results

Tablet Size	Dissolving Times (sec)		
	T1	T2	T3
Quarter	71	68	71
Pieces	47	48	46
Powder	24	22	26

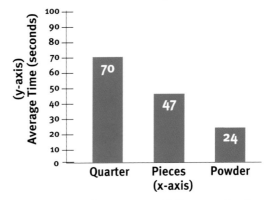

Conclusion

The results support the hypothesis. Increasing the surface area of the reactant increased the reaction rate in each instance. Overall, the reaction rate increased significantly each time the surface area increased. On average, the pieces reacted 33% faster than the quarters. On average, the powder reacted 66% faster than the quarters. Next time it might be interesting to see how water temperature also affects the reaction rate.

References

library.thinkquest.org, www.chem4kids.com

Glossary

catalyst
(KA-tuh-list) *noun* a substance that speeds up the rate of a chemical reaction but is not itself changed by the reaction (page 38)

chemical bond
(KEH-mih-kul BAHND) *noun* force of attraction between atoms (page 6)

chemical equation
(KEH-mih-kul ih-KWAY-zhun) *noun* a shorthand method of describing a chemical reaction using chemical formulas and symbols (page 22)

chemical formula
(KEH-mih-kul FOR-myuh-luh) *noun* a notation that shows the number and type of atoms in a compound (page 22)

chemical reaction
(KEH-mih-kul ree-AK-shun) *noun* an interaction between atoms and/or molecules that produces a substance with different physical and chemical properties (page 19)

coefficient
(koh-ee-FIH-shunt) *noun* a number written in front of a chemical symbol or formula (page 23)

combustion reaction
(kum-BUS-chun ree-AK-shun) *noun* a chemical reaction in which a substance combines with oxygen (page 30)

covalent bond
(koh-VAY-lent BAHND) *noun* force of attraction between two atoms that share electrons (page 12)

decomposition reaction
(dee-kahm-puh-ZIH-shun ree-AK-shun) *noun* a chemical reaction in which a single compound breaks down to form two or more simpler substances (page 28)

double-replacement reaction
(DUH-bul rih-PLASE-ment ree-AK-shun) *noun* a chemical reaction in which atoms of two compounds switch places to form new substances (page 29)

endothermic reaction
(en-duh-THER-mik ree-AK-shun) *noun* a chemical reaction that absorbs energy (page 33)

exothermic reaction	(ek-soh-THER-mik ree-AK-shun) *noun* a chemical reaction that releases energy (page 33)
inhibitor	(in-HIH-bih-ter) *noun* a substance that slows the rate of a chemical reaction (page 38)
ion	(I-un) *noun* a charged particle produced when a neutral atom gains or loses electrons (page 10)
ionic bond	(i-AH-nik BAHND) *noun* the force of attraction between negative and positive ions that results when electrons are transferred from one atom to another (page 11)
law of conservation of energy	(LAW UV kahn-ser-VAY-shun UV EH-ner-jee) *noun* the law that states that energy is neither created nor destroyed; it can only be converted from one form to another (page 34)
law of conservation of mass	(LAW UV kahn-ser-VAY-shun UV MAS) *noun* the law that states that matter is neither created nor destroyed during ordinary processes, only arranged into different forms (page 23)
metallic bond	(meh-TA-lik BAHND) *noun* the force of attraction between a positively charged metal ion and the electrons in a metal (page 13)
molecule	(MAH-leh-kyool) *noun* the smallest particle of a covalently bonded compound that has all the properties of that compound (page 12)
product	(PRAH-dukt) *noun* a substance that forms during a chemical reaction (page 19)
reactant	(ree-AK-tunt) *noun* a substance that enters into a chemical reaction (page 19)
single-replacement reaction	(SIN-gul rih-PLASE-ment ree-AK-shun) *noun* a chemical reaction in which one element takes the place of another element in a compound (page 29)
synthesis reaction	(SIN-theh-sis ree-AK-shun) *noun* a chemical reaction in which two or more substances combine to form a single compound (page 27)
valence electron	(VA-lents ih-LEK-trahn) *noun* an electron held in the outermost energy level of an atom; an electron involved in chemical bonding (page 8)

Index